DREAMPAD

DREAMPAD

JEFF LATOSIK

POEMS

McCLELLAND & STEWART

Copyright © 2018 by Jeff Latosik

All rights reserved. The use of any part of this publication reproduced, transmitted in any form or by any means, electronic, mechanical, photocopying, recording, or otherwise, or stored in a retrieval system, without the prior written consent of the publisher—or, in case of photocopying or other reprographic copying, a licence from the Canadian Copyright Licensing Agency—is an infringement of the copyright law.

Library and Archives Canada Cataloguing in Publication

Latosik, Jeff, author
Dreampad / Jeff Latosik.

Poems.
Issued in print and electronic formats.
ISBN 978-0-7710-7311-3 (softcover).—ISBN 978-0-7710-7312-0 (EPUB)

I. Title.

PS8623.A788D74 2018   C811'.6   C2017-904778-7
                                           C2017-904779-5

Published simultaneously in the United States of America by McClelland & Stewart, a division of Penguin Random House Canada, a Penguin Random House Company

Library of Congress Control Number is available upon request

ISBN: 978-0-7710-7311-3
ebook ISBN: 978-0-7710-7312-0

Typeset in Stempel Garamond by M&S, Toronto

Cover design: Rachel Cooper
Text design: Sean Tai and Rachel Cooper
Cover art: © Shutterstock, texture © Dreamstime.com

Printed and bound in Canada

McClelland & Stewart,
a division of Penguin Random House Canada Limited,
a Penguin Random House Company
www.penguinrandomhouse.ca

1 2 3 4 5   22 21 20 19 18

CONTENTS

1.

- 1   Dreampad
- 3   Permanent Indefinite
- 5   Holes
- 6   There Is a Delivery Specialist
- 8   The Internet
- 9   Komatsu Floodlight
- 11   Cats
- 13   Trans-Neptunian
- 15   Red Giant
- 16   The Home Checklist
- 17   Sky Pool
- 18   A Mile from the Bay of Biscay on Tour with Oneohtrix Point Never
- 20   The Eastern Massasaugas

2.

- 25   Troubleshoot
- 27   Centaur
- 28   Clearance Sales Are Adulthood
- 30   Swimmer
- 32   Spacetime
- 34   Silverado
- 35   The 3D Tour
- 36   The Fortune You Seek Is in Another Cookie
- 38   Life in the PhotoStream
- 40   And Missing Stephanie Stewart
- 42   The Bright Note
- 44   School
- 45   And I Looked Up into the Blue and Green of Nobody's Fields
- 46   Letter to Kyle Bobby Dunn

3.

51 I've Been Baron Munchausen
52 Growth
55 Cubewano
56 The Natural
58 Pop Rocks
59 Clear Giant
61 The Replay Review
63 Hidden Pockets in Parkas
64 The Good
66 The Connectome
68 The Surface Fuss
70 On Finding a Discarded Blind Cord Weight on the Street
71 The Adjunct
72 Osgood-Schlatter
74 Akasha
76 Pack

4.

81 Phone Booth Man
84 I Don't Want to Kill It, I Just Want It to Live
85 On the General Being of Lostness
87 Dream of Dee
88 Dryzmala's Wagon
90 Dear Listener
91 Platypus
93 Guitarist
96 The Joeys at Kangaroo Creek Farm
98 Only an Avenue
99 Oath of an Unaffiliated Boy Scout
101 The Great Illusion
104 Two Cells Made All of This
106 The Fly
108 Dreampad

111 *Notes on the Poems*
113 *Acknowledgements*

DREAMPAD

1.

DREAMPAD

It's this calendar I've dislodged and am playing
like a simple music grid controller.

It's the past, plus all I've sleep-talked
and confused with what took place

and it starts out with a pulse of light click-tracking
across time and space. I gather up some days

and make a living beat to layer over. Then the grid
populates as memory, which has reverb

and you best believe it has attack. Myself, age eight,
coming back from a vacation that my mother

and stepfather had themselves dreamed up
heading in the same direction for the last time

and I've got a salamander hidden in my hand.
I want to make a commune for the part-pond things

but when I look again it's just a smear of red
like I've wrenched down a nebula.

My stepfather looking out onto the highway
must have felt the same thing when he understood

my mother would be leaving—some general lack
over which the world comes tumbling again.

Hence, a trick I like to do. I make all that isn't
come to in a half-life of being dreamed and as I do the days

patch through in a way that's hard to damp and fade.
Strange, though, my remixing's not my stepfather getting clean,

or my mother finally getting to live beside the Atlantic.
I feel it in my hand sometimes, like a rubber band

has tightened in my wrist, but I play better than I once did
the older that I do. I missed something that made my life.

PERMANENT INDEFINITE

Mike, who lives in Paris now, tells me his work
has given him this designation,
PI for short. As if to bring up
some of the fine print that underwrites
even a synonym for *always*.

Our table in Nonna's Coffee doesn't have to be a table.
A salt shaker can show me where his workplace is
in relation to his apartment packet by Tuileries
and where the shooters came and tripped
the wire that let down a terrible nothing like a balloon drop.

It can also be how I tell him I'm still close by
to that St. George campus belfry where we did our masters,
where the art gallery's going now
and where I walk my part-lab part-something—
we're not sure because the kit broke in the mail.

I want to show him where they're putting
the boutique cubes set like aquariums on single
storeys so a passerby's seeing fills
the empty half of someone's life
so I put a finger down and say, *Here.*

It's always more common than I think—
intending one thing but with some possibles
floating there, scattered disc of all the essences
a meaning isn't but still needs.
I know some years will begin to rearrange us

as if they're explaining something difficult to each other.
Mike's a father now, too. No getting around
the knack or need for clarity to drop
down on its chopping block,
so I put a finger down again and say, *No, here.*

HOLES

A playground fight some years ago
years that haven't passed so much as they've stepped aside
and I came out the other end
with a hole in my trapezius.
Not like someone put it there, more like a space
was pushed aside and it appeared.

I learned to hide it. Walked with a hunch.
Whoever saw it did this little dance
of walking backwards and then defecting.
The doctor wouldn't remove it
because you can't subtract what isn't the case.
Fill it in and it remains.

Someone with a heap of Tag Heuer on his wrist
sits in his office and reaches deep
into the Cayman Islands. Meanwhile,
on the coastlines of Miami, Porsches sit
flooded to their dashboards. A hole is what your mind
falls into when a thing's hard to explain.

Sometimes I think of that rube Wile E.
falling through a bridge because the Road Runner
set one of those portable holes down on it.
How he threw it there exactly, none of us Wile E.s know.
I think all I've learned is that a hole's
not really there. That's how it grows.

THERE IS A DELIVERY SPECIALIST

We were standing on the outside of what was coming.
Trouble is, so was what was coming.

Conveniences gushed in until they were unintelligible.
As if we'd tried to avoid drowning by flailing

into deeper water. It used to be you could go a generation
and still know where your music was kept.

Now, my music seems to spy on me
and hide every time I turn around.

Car makers wanted to win each of us
by leaving nobody's wants unmet.

TV makers desired a papaya you could pick
right from the screen. Pen makers wanted

a pen you could twist so many times
you couldn't then find your way out of it.

One twist and this kind of heaven opens.
Some call it heaven. I call it a waiting room made of spares.

Still, the salesman's sometimes right:
it's nice to have your lights dim for thirty seconds

to give you a head start walking in the dark.
As if you needed it. Please, when it all goes out,

as in the trees, the sky, the house,

please

THE INTERNET

I first heard about it in a Burger King.
Its aims seemed as elusive as the stock ticker
or why some people stayed in marriages.
The future was flying cars, phone screens, and MiniDiscs.
I bussed tables with a cloth that mucked the laminate sheen
and, just that spring, an annular eclipse ringed the sky
like we were suddenly looking down a cabled conduit.

Then, as if an indigenous strain moving beyond a range map,
people started *getting it*, birdsong calling up from basements,
the pink noise, hiss, and crackle of a connection made.
And somebody already had some pictures: the body,
pixelated, bare, with the feeling you were overseeing it,
moving along the conveyer belt of banner ads.
Weeks disappeared as if dragged into a bin.

Somewhere, fibres tethered us to a warehouse or a factory,
but for then the feed seemed as ephemeral as a thought.
The search bar a mail slot you could lift
just enough to see inside somebody else's space.
It wasn't a place, but you could go there.
At night, blinds down, but windows open, flags of light
were quietly raised from main floors up into our rooms.

KOMATSU FLOODLIGHT
This floodlight is capable of illuminating large areas, even in the dark.
—Product label

Even in complete pitch-dark, otherwise known as me attempting "Helpless"
on ukulele made worse if others are there to hear it.

Even in regular darkness, which I confusingly think of as medium light.

Even in performance-enhancing dark, otherwise known as a Black Russian.

Even in that early evening dark we all must, eventually, sit down in.

Even in starlit dark, which makes, as Maxwell said, any person you're standing near somebody you love.

Even the dark of the matter I picture passing through me so weakly it could be my own half-educated imagining.

Even in city dark when what's passing now or whenever for God turns a tilt wand and everything five feet above you, up to a distance called the world, is illuminated.

Even the metaphorical dark I was in when I defended the group Ice Cream for Christians.

Even the dark when the deeper and more difficult day remains; after hours go slack and faces and paper stop raining in offices.

Even in the daydark! As that Riesling you held up in mid-afternoon in Chautauqua, New York, and said, *Look, a reverse lamp.*

Even in that sort of sociopathic, literal dark. Tiny reverse lamps for all.

CATS

That night I received a text from someone
I didn't know looking for Melinda and then—
I'd had the phone for four years at that point—
asking about her left breast's nipple
and whether or not it was tender still.
Tender. Such a word for a missive lobbed,
buzzer-beater faithful, into the black,
the way I imagine sending money out to a bookie
or a hothouse banking on a novel to cover their losses
and then my phone lit up with Gil's final message,
Are you still here? My thumb went still above
an autofill of *No*; a position can pixelate, too.
I imagined a doctor on the other end
in front of a chart or perhaps a concerned
but lascivious friend who'd gone off the grid
when, through a window, I felt the first true breeze of the season.
I noticed, then, no whiff of the abattoir
remembered it had closed down in September
all because of porcine diarrhea and condo developers,
two things that wouldn't have otherwise
explained each other, but in this case they did.
So, a brick-quiet building was what it was now
other than the cats, unknown numbers living in colonies
and eating the cat food some other unknown left.
And would they relocate or be freeze-dried in bins
or did they make do just finding food in a puddle?
You'd walk by and see cats shift in the branches
as if lynchpins in the leaves had come loose,
and some calico energy would quicken the day.
I'd gotten stuck here and couldn't uncramp,

so I closed the window and returned to the room
which was darker now, but it was that kind of darkness
that helps you see colour and, briefly, detail.
I laid down and listened as I often did, to the vents
as they exhaled endlessly. My phone was jettisoned
but still my thumb hovered over air, I think it was, air.

TRANS-NEPTUNIAN

Have you ever heard somebody read
their own poem on a podcast?
Every now and then it seems
they switch a word or phrase
to another that's not better,
but *equal* in all the ways you'd want to look.
So "If" becomes "I also like"
and "To deal with sensational loss"
becomes "When I get very high or low"
and so on—*

These subtle shifts of emphasis
almost seem to say the written one's
got somewhere that it has to be
and might not stay. But the read one
collapses back against centripetal force
and the poem goes on as a kind of quiet storm
bringing something into view:
greens you didn't think you knew
and blues that are no longer blue
as soon as you can point to them.

This remainder then: that it's as if
all this unravelling was as natural as saying, too.
And stopping there on a word, the read poem,
almost—and I should say *enjoyably*—
seems to skipper off to its own end.

---

* The dialogue in direct quotes is taken from Nick Laird reading his poem "Feel Free" on *The New Yorker* website.

Or maybe it just swerves when it was said,
Stand straight and just be read, my dear, my darling,
as if it wasn't that there were too many things
to name but *too many right words*
is another way of saying what I'm approaching.

RED GIANT

Composed of simple stuff that would make fluff
look Byzantine, it can never have enough
of itself and so it's screwed.

A night light, though, from eons back,
some core fuel burning like a childhood
in the black, and it grows and glows

and it consumes. Would that we knew
the kink in things that when you've burned
through all your credit you're just given more

but under different circumstances.
And how you might, unstuck from a lucky streak,
rummage through a closet as a red dress

falls across the bed like a solar flare.
When a giant grows, it's also said that narrow zones
once inhabitable become somehow temperate

as if where none were for life, now there are chances.
Look into one of those videos, concentric circles
of a supernova going off, and it becomes

a spotting scope, lens opening leaf by leaf,
and through it I swear it's possible to see
the one thing that you swore you couldn't do.

THE HOME CHECKLIST

Some general comingling of space
and location. Also, not sinking.
You laugh. But something set back
dare I say plain,
but spruce-able with small
and unobtrusive alterations.

Plenty of closet space.
That one is a must.
Fenced where needed, open where not.
You know, the general ethos
you could just see yourself
having a beer with?

Ranked school nearby but perhaps
out of earshot. No powerline static.
Minimal landscaping of the greenery required.
For that matter, no attic.
No tenant apartment
in which a family member might hunker.

All main lines wired. Fibrehood ready.
Gas leaks not common. All sockets modern.
Yes, if a deal can be struck
then we'll go over asking.
What was it, again,
that we were asking?

SKY POOL

> *My vision for the sky pool stemmed from a desire to push the boundaries in the capability of construction and engineering. I wanted to do something that had never been done before. The experience of the pool will be truly unique, it will feel like floating through the air in central London.*
> —Sean Mulryan, chairman and CEO of the Ballymore Group

One day it was there. A bridge you could swim.
And people you truly couldn't hear
doing their butterflies and backstroke laps
ten stories up. Oh, the glass floor must

have made a person feel like they were moonlighting
as an astronaut or had become half water-dwelling
even when a storm passed through and girders creaked.
And when the girl went under in the public pool

at street level and wouldn't breathe and all went
quiet as shadows do at a repair, somebody
was still swimming way, way up there.
So it seemed the arm span, turning,

was a soul loosed from this night and crawling
in the chlorine hue. When she was back, and the fright
that she had made declared, a wind did shake
the bistro sets as if there was, still, some spell we couldn't break.

A MILE FROM THE BAY OF BISCAY ON TOUR WITH ONEOHTRIX POINT NEVER

On a sleeper coach outside of Bayonne, by way of Utrecht,
one Daniel "Oneohtrix Point Never" Lopatin, opening act
for Nine Inch Nails, reclines in his cabin view and tweaks the buttons
on his Roland SP-555 as they light up and go dark with synths.
Which always makes him think of Bushwick with its bodegas
and flickering, its embarrassment of windows,
and doesn't then the transport I tuned up just to get him there
almost tip over a rock face in the South France air,
and with it towns and churches sitting in the hills like Frisbees sit
beyond their throwers, fenced out, lost, and all this
threatens to go down with him, but fear not, fair rider—
a belt fan's blown and with it smoke and coasting
and so there's time to kill. Ah, rural anemone but with no vehicle
or as the French say, *Comme il faut*. Let's say it's summer,
so he steps outside and hikes a road beside a field of bright canola.
Suddenly he's by himself, except he's got his sampler
and the sky's not breathing static yet and no window-plain glass
so nothing for the man who told him last year
outside of Saint Vitus that he was once a law professor at Rutgers,
and he knew most glass so well that he could ballpark
the money in a recycling bin down to the penny.
And so Lopatin sits down on a patch of grass
and rekindles a loop from a song he wouldn't say
and twisted to the side the sampler's pitch-shifted
to a building in the skyline lighting up its bright interiors.
So he's remembering buildings so clear with glass
they are their own reflections, like you could ring
each like a brandy snifter and hear the most pleasing electro pop.
But could you? No. And did the woman who once lived

two storeys up from his old stomping ground in Winthrop
and whom he saw bare-breasted once by accident
really disappear under a rift of snow deep in Colorado
despite what the palmist once said of having a son?
Sky shades to question, like hooks to sung refrain,
loopy little roads will highway out. But Lopatin's not mulling on this:
he's on a tour bus lane of thought that goes on unseen to me,
and to try to follow it is to end up in The Hub in Edinburgh
in front of the Roland Juno-60 "Judy" he's known for longer
than he's known anyone but his parents, and he's, hood on,
ripping up a set that has the whole audience lifting swing tops in
    appreciation.
Meanwhile, he turns to me and says, *Hey, man, what am I doing here,*
I'm due in Maastricht by sundown, or, at the very least
this is a concert and frankly you're interrupting. So,
I'm flattered but sit down, know better that the line is fixed
and it may be clever but you're going to have to live there.

THE EASTERN MASSASAUGAS

There are maybe two dozen left
in the Carolinian Zone, so one bad snow
or, hell, a heavy rain could put out
that last living flame of keratin.

Odd, then, that they're tattooed
in blots that look like hourglasses
as if they were the muscle
time sent out to collect its dues.

And true—should one sink its position
against health care-for-all into your shin,
you'll find there *is* a shortest route
to the anti-venom that costs

twenty thousand dollars for a vial and doesn't keep.
But I remember camp counsellors
in Tobermory used to carry axes
on their belts to behead the ones

appearing on the rocks to sun
leaving a mob of boys so quiet
it occurs now that they had
still only seen about five hundred Sundays

in their lives, had no words yet
for that strangely quiet stomping dance.
Perhaps there are no words.
So the Massasaugas are collected up

gravid female by female through sheer luck
or some circumstance we could undo
if we knew how. Once, up in Parry Sound,
I came across a brood of them sunning

and I've since asked my father how he kept his cool
with that sound like pushing sand straight
through the sun. *You know*, he said,
*they must be multiplying in your memory now
because that never happened.*

2.

TROUBLESHOOT

Eventually, you have to call. And you're met,
as ever, by the range of choices your qualm half fits,
a cache of wants crushed on a touchpad of options
that feel as though they've been free-floating

and present forever. Each selection
another wing of an office as every door
locks behind once it's closed and stairwells
are numerous. Some agreeable music, too,

if you knew to what you were agreeing,
and the effervescent pre-recorded voice
disembodied as the suiciders in Dante
tape-looping a script that only ever

accentuates the bits of silence stitching it together.
And what's the real trouble? You've gone off the grid,
the tether that bound you has slipped
and you've spelunked into a crevasse

of unseeing and not being tended.
Ho hum. Cue the scrum of sound bites
that collect there: *It just is what it is.*
But another inkling lingers,

one possible outcome on the other end not picking up,
and it flashes as a fever dream of speed and distance
collapsed, of clearing the barriers, of access,
and that general particulate: being saved.

Until support chimes in and absolves you from this,
and leads with that enthusiasm of the paid
but not-quite-secure in the position,
and initiates the standard steps of contrition.

What version is it? What make? Has a cable
been compromised? Have you practised in faith
that first basic tenet of the restart?
Conversation might linger here on the fact

it's summer somewhere not on your seaboard,
that your concern has been routed so far away
from its supposed place of known origin,
it stands to reason any ground gained here

will be more by fluke than by intention.
What's really wrong, though, is that you're still
hoping, maybe even caught yourself
knocking three times on the countertop

for the device to c'mon and just do what is honest:
cough up the stolen or the disappeared.
That you've been angered, or halted,
spoke with a trace sample of condescension

or slandered the company name—funny, perhaps,
how the smallest matter had within it a largeness
that couldn't be parcelled or packaged before
you were sent back out into the evening with discounts.

CENTAUR

One thing and plus another one
for good measure, thrown in,
as if somebody, Zeus-faced,
in a way it's hard to stop
imagining said, *Just one more
addendum or appendage*

until the lot spilled over into not
quite the thing it was exactly.
My GPS keeps getting closer.
Its voice more the kind
I'd just put on and listen to
like those noise machines set to heartbeat

or Summer Meadow. It's been said
that non-riding cultures saw the Minoans
and imagined centaurs in their midst.
Yesterday I double-took a GoPro
that turned out to be a lock of hair
and so I shook my looking in the breeze;

I think it almost fell from me for good.
Tonight, I meet you in a bar that's no more
than a hull of wood at sea somewhere,
and it's because we're there, I think,
that I'm really what I am, outside
of all the books that remember.

CLEARANCE SALES ARE ADULTHOOD
—after Glyn Maxwell

The most expensive thing is to live
on clearance sales—that's something no child
would wholly understand, but to me
it makes all the sense that pith could give.

You don't see it when you're then
and there. Not the shoddy quality,
which is a calculation you can make,
but the way a cleared thing will hang

the mind on it, and thinking falls and settles
where the dress was made—somewhere
that's just a word most days, placeholder
to a fullness that you can't fold up

or crowd into a shipping tube, or when the cardigan,
like a little bit of storm come loose,
hangs by the IV drip and the patient has
the fortune and the means but won't get better.

It's just another kind of reading is a way
I haven't quite explained it. So we put our backs
to this and pass the store, and what it stands for
is a kiddie pool that deflates.

This is a place where happiness will moth-wing
in for moments, and morning comes
so fast I'm still waiting for a long-passed one.
Adulthood isn't near. I thought I'd wake up one day

and say yes, I'm in it, if "it" is what they
all were saying would unfold. But no,
it's a thread I pulled once absent-
mindedly unravelling everything here.

SWIMMER
   —*after the Canadarm*

Deft space-appendage animating the blankest medium,
    picking through a serfage of solar arrays
        and repurposing

what's slipped from use. Like slo-mo replays of those
    shortstops who nab infield hits it clasps
        a payload

to its weight. And it goes on repairing what it is, which is
    a blinking in the blank, a swimming of a thing
        in briefness.

Would that it dipped an effector-tip into Earth-blue
    whipped up a supercell and sent it through
        azure Bahamas

or northwards where it was thought it couldn't go
    and kept it spun and glowing like new glass
        or a jettisoned thruster.

Might it Travis-pick the nimbus strands like an avid fan
    of John Fahey or dabble in stocks like
        the invisible hand

that steadies everything we can't see that cusps us.
    In this ubiquitous lack of air, all's clipped to what
        is possible

more loosely. And this arm can seem to stand in for the arms
    of everyone who ever reached for anything:
        Hominin

on a Saharan plain blinking through branches or a kid in Dali
    using all degrees of freedom to sweep for
        an x-drone

under a bed. And, then, just as easily, it stands for no one.
    It can press and curl the equivalent of a city bus
        and swivel

like a boom shot on a second take. On Earth, though,
    it can't even lift its own
        weight.

SPACETIME

As in the flour-and-water concoction
Polish migrants used to make while
they were given refuge in Uganda in '42.
You'd take the flash powder at the centre
of your life and stretch it out until
it seemed to defy physics.

That's how people ate. But it must
have also felt like you were spaghettifying, too,
so when the Bunyoro army hacked a clearing
from the elephant grass and served up borscht
to the displanted, spacetime maybe settled.
Spacetime—the thought that none of this is separate,

and what you've got with a time and place
is more or less a draft. Or, when you're on a raft
one puncture from the Aegean, the Caspian, or the Jordan
it's about how you can make a vessel
built for thirty transport double that
for sixty thousand extra euros.

Sunsets are still beautiful out there.
Minutes can stretch when you're anywhere
going back isn't. Stepping onto the truck,
the dinghy, the lorry, with patrol boats
ever circling in the distance
spacetime's counting kilometres then

by feeling how nauseous you are at the moment
or how a north star is the last knowable

thing but even that can loosen.
When Hashem was hiding in the bathroom of a train
the French National Guard poked around
in bags with the noses of their guns.

Then they were gone. He spent some minutes
that expanded outwards beyond a regular day,
the way he almost seemed to be
through the checkpoint, over the border,
in a tenement apartment overlooking a courtyard
in Stockholm through which people streamed.

SILVERADO

I swear that when I die I will wake up there
the vat my brain's been boiling in, twelve again,
plywood everything, a beige Berber carpet
and the sound of pines shook by a wind.

This would make it 1992; fires blaze on in Los Angeles
and I won't know where the cold war goes.
My sister's father is still alive in his creaking
mid-life-crisis leather. He's downstairs now

and the two of them are singing soul songs.
In this vision, I've been snagged by a thermometer,
its rising red, and I must stay here
away from school and the fields that are not fields now

and the sky is deepening with contrails.
There's a story my grandfather told about
waking up and seeing someone he'd never seen
before standing at the foot of his bed

but then he didn't say what happened after.
Who could? The rising red thins and tears;
whole years unflex whatever pose somebody
told them they needed to be holding.

And I'm unclenched. But what I am now can't
be made real to whoever was once lying there.
So I just run my fingers through the hair I had
until whoever's in the bed starts thinking of my sister's father.

THE 3D TOUR

I move through houses one by one
more first person shooter than the merely
passing through. Domestic life gone totally Escher.
Room by room in funhouse view,
until you're just back where you started

but this time knowing a little bit less.
There is a forest in Cordoba in the shape
of a steel top guitar. I pick it up and strum
a tune that spelunks through the steepest
of Sierras. I'm not sure how to put it down.

There is an island in a lake within an island
and yes, once more, within a lake in Northern Canada.
The natural world: you need an easement
just to build, which isn't exactly
things getting easier.

Up into the stratosphere, spinning through
the great and slowly gentrified of galaxies,
where all the event horizons slip down
into their zip-nothings to the voice
of Neil Degrasse Tyson.

Every fantasy I ever had would sound alarms endlessly
if I was there. Here's a planet where the air
might be enough for living things
but it's still too soon to call it. And then I get up
and put my apron on and check for my wallet.

THE FORTUNE YOU SEEK IS IN ANOTHER COOKIE
—Fortune from an actual cookie

Unless, of course, you seek no fortune,
and in that case it's here.
Just as if you seek no meaning,
you've come to the right place and if you do
it's in another poem, probably written by Kenneth Koch.
What you think should be is often in another life, not this one,
where the mail carrier sometimes pockets the odd letter.
That kind of thievery just shouldn't be—
so plain it's not fair to suggest otherwise,
but I will because I like fence cutters.
Lucretius thought the sun was the exact melon size it looked
which was so of-his-time it wasn't
and seems to call back only textbooks,
their broken spines flap-waving.
I was waving to someone the other day I thought was you
when you were actually in Fresno getting your doctorate in beetle sex.
It was rated topbest school but you were never happy there
and wondered why because look at all those beautiful rhododendrons.
Once, you said, *The life I seek's not in the microscope I swivel before it.*
Life was standing on a windsurf board in Hatteras, South Carolina
where I'm told the high beach houses willy skipper in the night.
You think you're about to ride a wave
that somehow turns the water white
as if blue was really the first right colour
but in fact the wave that's happening is just your own.
Standing on the highway flagging down roadside assistance
because the jack is in another trunk,
the itinerary's in another sense not even existent,
and you really wanted to go to Georgia, anyways—

and not as in peaches. Things go so fast sometimes they can't be caught.
Remember then when moving to label all the boxes well
because sometimes what you're looking for gets so lost
it might as well be in somebody else's house or life.
Let's say one night you were sorting through everything
that made you realize you weren't the person you had thought;
somehow you'd sliced through the thin adhesive strip
that separates each thing from where it should have stayed.
Perhaps you'd walk through every room watching sunlight
slow-tsunami the parquet with its lone blend of everything that is,
plus a cleaving quiet. And you might come to rest on a view
of somebody sitting on a stoop outside waiting for news
of a friend who's not now suddenly so far. Or even far-gone.

LIFE IN THE PHOTOSTREAM

There I am in Frankfurt holding a Jever
in a side profile shot that calls up
every family member and none.
I don't know where I thought
I was going. Those years are a vapour

and my photos all cascade in a stream
and are stored in what I'm told is a cloud.
What we try to keep becomes
a kind of condensation, doesn't it?
There are the photos I took inside the house

I was born in and something that hangs
over it went on un-captured. Suggestions
to rename it photo-storm or photo-cane
haven't quite stuck. Whoever I say it to
can't say it back. There's a photo of friends

waiting for rain to let up or somebody to enter;
a kind of forced angle that's stretched
out someone whom I've lost touch with
into a homunculus that's all nose
and fingers. It's enough to make you think

that what once took place is a sea you can't fare.
Here's one from a plane window,
the sky but from the other side — clouds set
on a permanent pause so still and wide
they seem to be a continental shelf.

They define a possibility without revealing
anything. There—that one of my grandfather
singing that song he sang again and again
beyond the point where he could tell us
(and none of us had thought to ask him) what it was.

AND MISSING STEPHANIE STEWART

One morning she didn't call in her report
from the fire tower ten clicks outside of Hinton
so her supervisor drove in and found
a lone pot with its water boiled.

Firewatcher, she could spot a plume of smoke
in all the places that you couldn't.
It was quiltwork she took into darkness
and, at seventy, improbably got by in ways that no one saw.

That kind of seeing's what we're missing—
and likely for good. The pillowcases, bed sheets,
and Navajo blanket left behind are simple
and yet massive as an un-ticked box of sky.

The cash reward remains unclaimed.
Search teams traversed five thousand square kilometres
rife with all those combustible kinds of spruce—
angled through that precarious splendour

a person must be careful not to watercolour in his head.
Found nothing more than what they knew.
There are many things to mess up in recounting
what took place. The government, known to burn

seedlings to spare the larger blaze,
has since adopted safety measures
including fences and two-way radios with panic buttons.
In the place of everything we never saw

that happened comes the world. It rolls
in with a weather, is marked by a date,
is one thing until it's another or two things
and there's too much of you to see it.

THE BRIGHT NOTE

> *No silver linings and no lemonade. The elevator only goes down. The bright note is that the elevator will, at some point, stop.*
> —Douglas Coupland

It's hard to step out of a dream. The world
can be so boho, but it can also go all doorknobs
on you. Those days the numerous gone
still rise from us, like breath in cold.
Only a small, wet gland truly gets that they're not here;
all the rest is very strictly on a need-to-know.

There's an elevator travelling down.
No silver linings and no lemonade. For longer
than we might have thought possible.
It keeps passing the floors we thought
all these stunt falls were breaking to. But it stops.
The other day I was waiting in the documents office

for a form that proved I was myself
and above me a hurricane wheeled across an HD screen.
I was thinking of my friend Sam sitting down
in Volcanoes National Park and watching
a pack of silverbacks process leafwork so thick
that brushing it aside only made the pathway worse.

To see the apes' plush strength contemplating
something deeper in the green than he could see
was also to be rocked. And through the leaves that moved
like taking yes and no and making them one nod,
he saw the troop of guards carrying Kalashnikovs
protecting all the life in the nearby vicinity.

Sometimes it seems no stretch to wonder
if the world one day may just yell *Cut!*
And send the trillion key grips of reality scrambling;
and spike the microphones in the grass
so no one sings; and spin again the giant carousel
I must step off to just see anything.

Every now and then, I kick a rock to know
I can't live every day in the temperate zone
around the temple that (it strikes me now)
is maybe the best synonym for mind.
Then they called out my number
underneath the hurricane and I straightened up.

I had no slips to prove who it was I was.
Sometimes you're really just hanging
on an umbilical rope. I don't know why
the slack line tightens, but it does. I don't know why
a figure eight that holds you above emptiness
is a clef for all the brightest notes.

SCHOOL

Everybody asked me what I already knew
and that's how I learned. I learned in books
I gave right back or sold and didn't even crack
up when the joke went round like pepper spray
in a far-off row. And even this was a test.

School. It's measuring something, no one
can say what. The teachers do, but it changes.
They park their cars or come in on a bus
and make a fortress of a book or gape
out of a window and then nod their heads.

Jobs bend us in them and have a gravity
from which only money escapes.
We look down into phones as if from a tall building.
They buzz on like we keep finding the edges
of some invisible electric fence.

The school used to be better or worse, we're not quite sure,
but we know it did change, and there isn't a place
that you can't buy a coffee and our sports team
is doing well at something that involves a mallet.
It sprung up close to where I lived, so I went.

Like many things, it used to be a factory. It made corn chips,
so there's corn chips everywhere, corn chips and coffee.
It filled up with buildings, then we came. Classes filled
and soon people were spilling out into halls.
I spilled there, too. Then, I was standing outside of it.

AND I LOOKED UP INTO THE BLUE AND GREEN OF NOBODY'S FIELDS

What it's like to be one of those common tricoloured bats
that will slip in through any spare flap a house might have
and make its slant way to the kitchen sink
to free up liquidity and no doubt send the shutters
flying open in whoever finds it first,
a bit of dark going out on its own interests—

is like stepping into one of those revolving doors you find
in public spaces as if you've walked into the paddle wheel
of a boat, but instead of taking you to Wachau Valley
or rocky grandeur it just spits you back out
again into the foyer of the Two Seasons
as if just staying in one place required an engine.

If I could be anything I'd be a ruminant grazing on
a little bluestem in Tahoma holding down the grassway
as if that, too, might up and disperse outwards
to reveal so many roads there could
never truly be a way. I'd be one who kicks up
seed and prunes back the shrubs

that would become invasive trees.
If there is a divide between us and it can be crossed,
then I will. If not, I'll be here at whatever
little hill I mistook for too much. At night
sometimes the sky stays blue. And should you ever
need me, I can carry you across a river.

LETTER TO KYLE BOBBY DUNN
MUERTE
—Graffiti on 158 Sterling Rd.

I wanted to say I'm listening again
to your *Fragments and Compositions*
and the one I've played a hundred times
is "Tout Voyeurs." I've counted the few tones it is
so many times an endlessness has opened,
and I'm wondering how a life in listening led to this:
drone, ambient, minimal core. You hear of people
wandering into a town they'd never
turn toward while driving by
and then one day up and living there.
Not the pound and surf of metal I could have waded back to—
four notes and just the space of having nothing there
but any space a mind will fill, and still those tones
can be the time you wake up half-cut in your dream.
So I'm getting old and getting less,
but you are young and did your stint in Brooklyn
where some sceney publications even profiled you
and there were living rooms and churches
to unfold chairs in and find acoustics.
Now when I pass the demoed building
by the Nestle factory where I live, I picture you establishing
your small settlement of pedals there—
your songs could ivy up the trellises of windows,
yawning and breathing dust, slowly turning on
the lathes and blades that spun there once or the dishwashers
and elevator shifts and thumps the building will become.
I used to have this thought that one day
I'd wake up in a different life, and all *here* was

was an atomic flash. Some nightmare of a self:
one long, unbearable sustain, a noise that funnelled up
and fell on its own weight but couldn't scatter.
If you pare back any song to a constituent tune,
what lingers when the counter-point
and contra-puntal flicker back is a seemingness of presence.
Like a soul, just what was there the whole time,
really, and so it's no stretch then to say it's always been,
a transparency or purity of being. Then, I'm suddenly seeing
those glass rooms being stacked against the sky
and how sometimes they all teeter like a game of Jenga.
I was just this tumbling guy who went on through
the many living rooms of the only-ever-really-rented.
I just wanted to hold these things together: the beauty in your tones
and then the wasteland that a being could one day become.
Something still knocks around in me, I admit,
an elevator lift of what used to be dread;
not soul but where soul was once concocted.
Surprising how long it stuck and hung to me,
so just let me say it: there is no other life to wake up to.
Did you hear of how they found a new species
of hominin in a cave called Empire
in the Malmani Dolomites? People slid down
a narrow chute of stalactites and found the bones;
Homo naledi, which in the Sotho language means "star."
They lived three million years ago, buried their dead,
and by their bones we can detect a resonance,
not *MUERTE* but the elbow room of the universe.

3.

I'VE BEEN BARON MUNCHAUSEN

Whenever it's clear I might be firing on some cylinders,
doing another's little jig of *good job* or *all is well*,
I've also felt the end there ribboned in it like a muscle.

Whenever I've seen the money I've spent and tried
not spending I've been pelted with the thought
of it travelling at the fastest speed we've measured.

Whenever it rains days on me or it dries up
and somebody's beautiful commencement speech just will not end,
I picture all that rain again and I can begin to smile.

Whenever I've heard reality's only just so there,
the tracks don't meet, a body's mostly just a host,
I lean back in my chair and a breeze enhances what I'm feeling.

Whenever I see the ones who left me holding the hundred-dollar flowers,
with their houses and their kids all flying by me single file
I can see the storms unpacking like machines off in the distance.

Whenever I'm braced against the quicksand or the ocean
or the audit, ready to tae-bo alligators if it all defies my druthers,
I can hold myself up with all the lifelines I no longer have
and say, without any sadness, none of my dream homes wanted me.

GROWTH

The *o* in it lasts for the time in a blink
but it's tied off with a clean shovel tap
of consonant as if something there might
swallow up all time and sense. *Growth.*

Fish blink off in a river that was once
a lake that was once an ocean.
They are slow fireworks for no holiday,
more gone for being captured in a GIF.

Meanwhile, my hand's locked
on Channel Everything so that I forget
that it's *my* index finger scrolling.
My own hand's the thing I'm battling.

And so whatever lumps appear I disclose;
whatever minute specs awake like animated
full stops and float in grout I dutifully
try to trap and kill. Twin currents—

to grow and to stop growth
and still the word seems to open up
a space around which all manner
of straight and crooked thing must be erected.

I picture a river around which leaves
have framed a view you might write home about
but you don't because that would letter out a state
it's put you in, a happiness.

Lost so easily it's as if it wasn't owned,
not growth but the moment before
the saying. The city I used to live in has grown.
In the ruined factories, there are weddings.

It was all Ojibwa land once and before that,
the slowest curtain of a glaciation.
I have a daughter I don't have
with not a resolution but a blankness

for a face or rather a face that changes.
She doesn't grow or I grow around her
and she visits all the more.
Peace is what I tell her blankness is.

What she knows is what rearranges
itself for a reason I can't seem to guess.
Growth. There's something it's all spinning
on that can't be said. Or if it can

it will be right by accident the way Epicurus
was about atoms swerving in the void—
the quantum in a coconut shell.
Down it drops or up life pokes its baleful

tortoise head from it someone should say
and if they haven't then I'll say it.
Once, in Bangkok a tuk-tuk driver stopped in mid-day traffic
and knew the locus points and fissures necessary

to crack a coconut I'd purchased with one tap.
He was gone before he'd heard my thanks.
Inside, the water was so cold that I still
think about it when I hold my finger

under a faucet, feeling for a sensation
I'm not sure I'd know I've got.
And then in a cab on Peter St. the driver motioned
to a screen on his back-seat headrest

and I tapped my card like I was never
even really there. Beyond that the only thing
exchanged was pleasantries. Don't worry,
tech support will tether us back to all

that can be billed if not leaned on well.
But there's forever an addendum here: error
is the loam from which layoff, retraction, bust
and frustration bloom. And us.

CUBEWANO
—*Pronounced "Cube-wawn-oh," a nickname for QB1, the first*
trans-Neptunian object discovered in the Kuiper belt

When David Jewitt looked into his lens
on Mauna Kea one night in 1992 he saw it—
lone outlier, as if from a game
that had been packed up and long won

and what remained was just this lone Hail Mary
that never really got squared.
Not among the sleek and smooth-seeming
entourages of moons and planets;

unconcerned with the inside politics of orbit.
It sped by in what we can't now call
only just the empty dark.
I wonder if he saw the smashed abacus

of light and blinked away a simple math.
Maybe he asked if moving forward really just meant
staying put to see what's only ever been present
in a kind of not-plain view. Last night I spent hours looking

at the little needle moving on my screen
predicting which way the vote or the weather would go.
I thought: I never would have guessed
I'd love someone who couldn't fall apart

in a FreshCo beside the many gluten-free cakes.
And about ten years, I'd add, since I sat down
and wrote my first what I would call
(because I lived all that time that happened after) poem.

THE NATURAL

So Yaadi Tremen pitched half a season
and was out with plantar fasciitis.
It wasn't fair but then there was *the talent*,
which is what Dr. Rory Keane who performed
the release procedure had in spades.
He poked around the heel spur
with its bird beak and its cresting wave
and even managed to daydream himself
carving a shorey break in Biarritz or Shonan.
Meanwhile, Rory Keane Jr. (who couldn't tell
blade from whisk) was slowly accepting the idea
that he wouldn't be the governor of Maine
much less the mayor of Newton,
a bell curve of fence slats that, should he be stoned,
expanded like the rib cage of a humpback
as he drove home avoiding all lit roads.
He'd say, *You'd do the same* but not to anyone,
much less his father who lived his life
in a percentile Rory Jr. thought of as a kind of summit.
Hence, the trip to Mount Katahdin
on which Thoreau stood atop in 1846
wanting to see the soul of nature and then
said, *This was that Earth of which we have heard,
made out of Chaos and Old Night.*
Quite exact, despite when you think of how
a doctor can elongate a calf muscle
quietly without breaking much of a sweat.
Or how Tremen once threw 100 mph
with a chest cold that had sapped him
almost to the point where he would say

the natural is too expensive.
When we think of the place we're meant to be,
doesn't it all seem to unfold so effortlessly?
Which doesn't get Rory Jr. coming back
and thinking to himself that he's avoided lit roads all his life—
and like that he's skidding on some ice.
When he recovers, it will go unseen,
giving him the time to sit and stare into the darkness
once again but this time think, *A chimera almost killed me.*

POP ROCKS

We had this game. You stood and shook them into your mouth.
Then, Coke. Then pain.
If we could, we counted, *One, two.*
And then we spit when it was too much.

You wouldn't have known the city then.
It had this narrow, empty horizon.
The buildings now are architecturally twisted
like they're being half-nelsoned for cash.

You did it again. *Three, four.* The circle grew
and grew and took more in as we learned
cell life started. Mack Johnson held it in for
five minutes. Then he promptly, silently left.

The city in exploded view. Cranes, more cranes,
two bedrooms, three. Skyline almost fizzing
like those long holds where the spit was blue.
I have no deeper story to tell:
There's an opposite to singing.

CLEAR GIANT

Imagine you do something small. . . . So if I forge a cheque . . . minor offence . . . and in the process of arresting and convicting me maybe the courts do a brain scan and they determine that actually she has a predilection for far more serious crimes. . . . She hasn't done any of those things yet she could commit a violent crime . . . if all you're worried about is what she might do, then you can keep her in prison forever.
—Nita Farahany and Jad Abumrad, "Forget about Blame?" episode, Radiolab

Here—not prison. Not really the occupancy
professionals state. Just a clear giant,
an optical cloak of steel and glass keeping us.
We *live* here, they say. We say it, too.

Except some days we're more, *It really just is.*
So we do the things we would have done:
water the hostas, mend a torn latch.
And on evenings when our zone plays

slow pitch long into dusk, then we field.
We love one another until it's too much
and Jerry or Avi's out sprinting again
beyond the gate, reaching for the clear giant.

How long has it been? I never wasn't.
A machine just glided over me like the angels in our books.
Some nights they show a PSA. Once, an aerial shot
of the brain could explain away a case or two.

But when we dropped down to city level
via fMRI and then GN3Q a tiny bit of fluff
or damage became the devil in the details.
What to do? Some nights I just sit and look

at my hands. In some whirring vector machine,
I'm dinging red, a cloud of killer I can't diffuse.
It's not human to not see the walls, Avi says
and sprints again. Or it is. Some nights I take my walks out

as far as I can go without the thought of returning
to my notches on a lone tree counting all of the years.
A room with no chairs and a view of a town
appears like a beacon whenever I close my eyes.

THE REPLAY REVIEW

It was a new challenge about the place
one thing ended and another began.
In the stands we watched and then
rained tall cans down from the blue
as if forever was just a vendor.

It was about a line in the sand
somebody had up and called blue ribbon.
A supposed given that was more a command
so there was a tribunal and then a long deliberation.
There was a common conclusion once.

Then a bunch of crummy pamphlets.
It was a beaut in one kind of way.
A territorial dispute of whether the fence-post
was foul or fair in a deafening boo.
It was everything peer-reviewed

up in the air set to blaring country
music and each citizen elbow-deep
in their data plan and their to-dos.
It was waiting. It was hard.
It was discovering that our camera-flipped

phones when turned to each other
created a kind of infinity mirror
making the whole scene more fun house
than a place you'd ever want to keep score
of anything true. It was a hunch that reality,

never more tricky, kept moving quietly
in and out of view as if stuck on one of those terrible
hot dog carousels. It was being so lost
in the inside baseball and the legalese
we couldn't tell the storms from the breeze

and couldn't freeze the bobbleheads
some other team, and then our team,
were becoming. Man, it was really bumming
me out. I was in the nosebleeds
wondering if I might just up and blow away.

I was watching all of us hovering there.
It was all hovering. A kind of slow flash
and it moved like knees do when the jury,
the crew chief, the judge, the worry of doctors
and the sea of committees are taking their sweet time.

It was arguing about that initial challenge
or at what point the call had been made.
No tape on that, though. All the times
I gave up on the final one I came back less afraid.

HIDDEN POCKETS IN PARKAS

Sometimes I like to sit and slide
my hand into one. The Sub Zero line can feel
like an alpine summit in itself.

The padding is called Thermaflex
which is really eiderdown in the way
that an impression of Christopher Walken

is the man who once played Frank Abagnale.
It's no big glitch to just start losing track
of what's really apparent—

but let's say that the lightest twitch
makes a tear in the warp and woof
a salesman swore on his own brother

was foolproof. And so suddenly I'm cut loose
from the lining that was meant to shield me
from whether I was really in the Eaton Centre.

I wasn't. I was walking through a field
somewhere in Hindu Kush.
I was kneading a tiny hit of hash

and combing a helmet into my head.
A sky chewed up by Chinook blade.
My smartphone was rifle butt.

And so many hidden pockets I couldn't
remember where I'd put the most identifying
of my documents: a letter from my stepmother.

THE GOOD

Oh, floating good of just above eye level, of non-blinding rays,
a bit taller still than anyone who bristles
at attention for the anthem or the silence—
let me remember the many Japanese women
who filled their breasts with silicone to please American recruits.
Many of them died, oh good of the high up in a tree
with no branches for the first ten feet then *boom*
it all comes flailing its arms for help.

There's an ever-shifting line we live by trying to walk
holding all the mismatched things we've sworn are not.
They won't settle because we're each other's answers
but we're separate. Yesterday, an officer wouldn't
acknowledge my reports that someone by the slide
was deep in trouble, and when I looked back, dear good
of the hundred hatha poses on a single ankle,
he became a tree himself, which is to say he left.

Sometimes you run away or stay too long,
have to live the hundredth day wiping gravy from the platter.
Sometimes you hang around kite level, unable to be pulled in
well enough or winds will whip you higher
than a human eye can see. Otherwise, you're not unlike
a favourite band touring Cleveland for one night
but all the flights there are being diverted
or delayed. I wish I knew you better

on those days when it seems the leaves are just these bad
magician hanky tricks. We have to postulate a fiction.
I could follow what is real down to its furnaces.

But I do see you out somewhere that Google Maps won't
guess the driving time to accurately. Somebody's limbs
are being switched off like lights in a room that's being left.
It's almost empty now except the last light whose glare
has now turned to soft candle glow at night. You're there.

THE CONNECTOME

When they laid the neurolink down on me softly
implanting the plugs of everything into
my late stage of limbic thinning
all knowledge moved through me as breath might move

in a napper. Some smell to a crocus, true,
but no crocuses seeming the slowest applause.
No caring that they had an ache-smell
that welled itself downwards, were more smell

than anything. When they peeled the server farm
off me as if it were clothes long drenched
in a storm that had lain itself down,
exhausted, I slept.

This one shored up memory
of Kyle Wallace in grade 3
aiming his air rifle at me one Sunday
and then whispering, *Bang.* I woke up to that.

I reached out through eons of air
and tried to fasten his childhood back on him
like ice that had fallen
from a high Artic shelf.

I can know property, degree, and position,
but it's very hard to follow meltwater.
The connectome could only duplicate me
to a pulse that could live on its own energy, forever.

Thank you, he said to me one night
after all those eternities that turned out to be years.
We were under a sky so punctured, it was bright.
I was looking upwards at my being gone.

THE SURFACE FUSS
—for Natalia Molchanova

The "surface fuss" is a term Molchanova created in her practising a technique used in extreme sports called attention deconcentration, where an athlete's attention is paid to peripheral and seemingly mundane body functions in order to endure repetitive or dangerous activities. Molchanova, who went missing during a recreational free dive in 2012, is considered to be the greatest free diver in history.

The surface fuss was everyone that stared
into Balearic blue waiting for her to resurface.
She hadn't really gone that far this time,
just another solo tryst without a spotter.

It was the stopwatch ticking and the oxygen
in the holding tanks, a simple hell for the marine life
she emulated in her monofin. It was the lanyard
she unclipped to move out of the light.

Hammerheads circled above her in the summer,
but at a hundred and fifty feet the surface fuss
was them, too. She was known to loiter
just below the average depth of the English Channel—

so the surface fuss was also channels, rivers,
inlets, and their levees and dams.
It was the Russian Property Developer
who was paying her for lessons.

Certainly his super yacht, which had a small raft
that they'd taken two miles out to clock times.
It was his shoreside home in Ibiza.
It was patrol boats and the entire Spanish Coastguard.

In two days, it would be a robot coldly looking
for her human form. It was dream stuff.
Capote, played by Philip Seymour Hoffman
said: *It's as if Perry and I grew up in the same house.*

*And one day he stood up and went out the back door
and I went out the front.* There was no place, perhaps,
the surface fuss wasn't—everything
that held us to the true like a metal column to its weld

when it had no reason to. It must have been
that she could shake the surface fuss off better
down where it was only breathing, heat,
and one's wriggling not to become another sunken thing.

It was pure luck that buffered me that day
I went under in a tiny lake in Wiarton, Ontario,
when a current pulled at me
like a loneliness lived down there.

Coming up, I remember saying to the sky, *Don't go.
I have a tag that shows I dove the deepest anyone's dove.
Have some air waiting there for me.* The surface fuss
was why I should have come up when she didn't.

I did, breathing air and water for a minute
or an hour while my mother and somebody else's mother
were swimming out to me, though it was tough
to tell one from the other.

ON FINDING A DISCARDED BLIND CORD WEIGHT ON THE STREET

I know *now* it was a blind cord weight somebody had tossed
but back then it might as well have been plutonium.
The steel bar clunked out of its casing.
An unmarked currency—not meaning, yet,
because we hadn't done anything with it.

Someone said it could be priceless. So in my head
I started buying all the things I didn't really want.
Someone used it as a dowsing rod unsure
of what to find under the fertile townhome lawns.
Tempted now to frame it as the I in its hard opaqueness

there on the curbside looking like trouble.
Every now and then I still pick it up and it still is.
Always out of its element, reflecting light in ways
that make you squint, it's a minus sign subtracting details
from the mix until all that's left is this one detail,

this nadir that I've never been further from
as when I'm holding right in front of me.
It's always colder than I'd remembered,
always just a little smaller than I minded.
So very small. There's still time.

THE ADJUNCT

Forever in the far row searching for the Wi-Fi.
First let go at the hint of a bust.
Where the jobs flaked and blew away
leaving the truer or the ones said to be true.
That tomorrow's okay but the one after that sort of trembles, leaf-like.

The contract zone. Where enrolment fills in
and the quarterly finds its profits.
In the factories that smell like scorched honey.
Dropping through the trap door of test scores.
The extras who will have to clean the stupid party up.

Shuttled in a friend's ride or labyrinth of connections
beside the bruised farmland and industrial parks
the adjunct's always sat beside.
The geniuses of politeness who are never late.
Flinging some filament anyone living will recognize and say, *Yep,
    been there.*

I've been flush and been so far from square I once lied
when exchanging insurance information.
Maybe our dreams are just these hot potatoes
we keep passing on. I know because there are no hit songs about it.
So was I not enough or too much of something?

Was I ever going to be the one who couldn't handle
his drunk? You can worry about where it is you've gone.
You can pop your gum, expanding and contracting
as if time is passing quickly somewhere. From the bleachers
of this evening's air even losing the way is adjunct.

OSGOOD-SCHLATTER

Little bump beneath my knee
I run a finger over and remember you
once making ice flakes with your hockey stops
in Nathan Phillips Square. Then you
were suddenly horizontal in the air, crashing
down like all the time until this moment.

My leg was what was there between you
and the Earth. Now you're gone and this injury's
just some evidence you were, I guess;
fascicled with memories of you picking up
my sister with your hair newly frosted—
the kind of change that never quite

inspired confidence. Then you became so thin
it was as if you fled yourself and left a consultant to explain.
Now rain is adding visibly to puddles
which, despite the din, don't become seas
though it seems they might. The gravel trucks forever go.
And, yes, sometimes when I'm just reaching

for the Hellmann's there are seconds
where I still think you're knee-deep
at the shorelines of Key Largo.
You liked it there, far away.
I see you with a cocktail umbrella
held up to the sun and know halfway

to anything's akin to being in some kind
of trouble. So I go over it again, this little lump

of once-active ache. Come to think,
it's more like a second Adam's apple.
It's not that I have something to say.
It's that an ailment made a voice.

AKASHA
—after Alice Coltrane

Turiyasangitananda was her name.
In '82 she recorded *Turiya Sings*
in Malibu, an album I can't do without.

John had gone by then. But in Akasha,
all his days were filed and stacked
in a cabinet where the loved are kept.

Turiyasangitananda was her name.
There is tuning in the timpani
of upstairs tenants, a quantizing in the coral,

those little riffs flung through the rocks.
She knows this, and when she plays
it seems even a doorjamb shimmers

and when she doesn't, even sitting still,
it still seems somehow as if she is.
She went a way no one followed well.

Turiyasangitananda was her name.
Turiya Sings is now only a digital rip,
a cassette that's up and gone to heaven

with its never-ending streams.
The filing cabinet of the universe is too deep now.
You'd need another universe to open it.

But there is another. The stars all
string the darkness as they flee from us,
but the stars don't play like her.

PACK

One night five hornets alit at my window,
white but dipped in black, or black but dipped
in white. One got in through the screen,
and then it got into the ceiling lamp
through some slit or break I didn't know was there.

Picture me with a spatula thinking: *I've got to get this over with.*
I'd take some steps toward the soft
energy-saving glow, and then as if a live wire
had come loose, the trapped charge
sent me flailing back. Back and forth, like that,

the flame sound flickering while I'm sitting *here*—
three years on, wondering what they were doing
five hornets just flying, probably not in a V,
but together. It's like that flicker
was a living asterisk. There's a memory

that's set its almost fever-inducing sac in me:
a kid who stepped on one of those hives they build
on dunes and then got full-on swarmed,
stung so many times his mother, a nurse,
just stopped pulling out the stingers and said, *Rest.*

But the blur of him flailing in the nameless
river some period of my life backed onto hasn't left.
An addition: some googling doesn't give me much
to work with. The European Hornet, for instance,
has been known to hunt at night.

Some wasps are defined as solitary, and, in fact,
a hornet is a wasp but to detect the difference
you need a magnifying glass. All this useful
in its way. I can hear that buzz
in the lamp, which had the curvature

scientists have since concluded the universe
doesn't. Stuck in there, the socket's buzzing
held the hornet's with its own until it became one buzz.
I know I got it. Which is to say I know I did
what was needed, as it is needed.

The thing is, this back-and-forthing leaves
something in the soft lamp of my life:
I'm old, so fast, the hornet's still there, buzzing.
The people who've been sitting around me have taken
their break. There's laughter somewhere.
Some part of me comes loose and floats

up to the ceiling and I see the beautiful legs
wriggling, flailing, lifting. The way rain moves,
how my mother used to wave to me, those nights,
in the cold when I was finished karate
and needed to be picked up. Then I'm picked up
and carried over none of the places I know.

4.

PHONE BOOTH MAN
   —*Credit for actor Michael Pecina in* Out for Justice

Hard to know exactly when he started training at our gym.
It was as if we held up a UV light and he walked out
of the ambient hues from a darkness
that we didn't know was there
and couldn't find if we wanted.

He was memorable for his odd introduction.
"*I was Aikidoed into a phone booth by Steven Seagal
on the set of the movie* Out for Justice.
*And when somebody finally opened it
twenty years had passed, and here I am.*"

I'd hold pads for him. No pop to his punches.
No torque on his kicks. I had to fake the force
of his ad hoc Muay Thai. He had a general look of farm strength,
but something was not quite firing its cylinder.
There was something wrong at the centre.

One day, he approached me. *I saw that wrist lock.
That's Aikido type stuff?* He was panting.
Sweat stains were living shadows on him.
I learned some in college, I told him
while unwrapping my hands.

When I sparred with Phone Booth Man I felt myself
suddenly dodging and feinting, tooling him up
as if he'd suddenly been slowed down on tape in front of me
and I could pick out smaller slices of time.
I told him: *Stay down.*

Months passed. Then one day he told me
Jimmy's Corner was putting a plaque up in his honour.
He said there'd be a government donation and Seagal
would even make an appearance. So I went. The owner joked
that Phone Booth Man could now make all the collect calls he wanted.

But Seagal was recording a fusion rock album in Milan.
The local reporter hit on a waitress. A waiter spoke
about his surprise when Phone Booth Man
finally exited. *"I assumed it was a dummy
the production company left. It was a person . . ."*

Ah, Phone Booth Man was training for Seagal,
I remember thinking. I watched his odd and stammering acceptance
of a novelty cheque. Then people asked him
what it was like in there all that time while he sat
by the jukebox blaring rock ballads.

Now more years passed. A ligament tear straight ensnared me.
One semi-professional fight to my name,
to a scrambler simply named Lights Out.
There were screws in my proximal phalange
as if I'd been wedded to a big factory that only made gears.

Phone Booth Man called me. We met at the place
we met at staring into the options.
*"People ask me about that lost time, and the only way
to explain it is like when you lose track of a day
then look at your watch. Well think—years."*

He was aging fast. As if his body was paying interest
for what had gone down. Like he was spending
all of his hours on credit. *"Look at this botched Botox*

*and these bad plugs. All that didn't change much.
Look at the way my face seems to be saying*

the good beer's been wanted and it's been drunk."
He paused for what seemed like a Holocene of minutes.
"*I want you to hip-toss me back into the booth.
I know you can do it because you did it before.
It needs to be strong—a good clean throw.*

*You think any of these jabronis can do it?
You think they know shit?"* When he said that I felt
a pain in my knee spread upwards and downwards
in equal measure. There were blots in his eye
revolving like planets.

He waited while, in the restroom, I watched the skin
on my knuckles flap under the blow dry.
Like one of those YouTube clips slowed down
of some KO that makes a person seem more water than flesh
more wave than particulate—

I opened and closed my hand. A feeling came
as if up from the darkness of a glove. Back outside
I sat while fried rice rained down and all the time
that had passed since I first met him in his cut-up T.
I squinted and my head was tilting suddenly.
I was actually seeing Phone Booth Man.

I DON'T WANT TO KILL IT, I JUST WANT IT TO LIVE

I've been trying to write "Those Winter Sundays"
for fifteen years. Robert Hayden was nearsighted
and grew up in Paradise Valley, Detroit
as if you just do. Banked fires have blazed for me
often when there's no fire anywhere.
That's one thing a poem can do.

Paul Farley's "A Tunnel" is another one I've been
trying to get to for it seems half my life.
He once experienced a condition he coined
"falling up syndrome," not trusting his feet
to stay on the ground. He was twenty-one then; I've stepped
onto his *Brighton* line for the one thousandth time.

Hayden died in 1980 and Farley lives in Lancashire.
If someone aimed a gun at either I'd likely jump
in the way. I guess that's three things now.
Time is holding Hayden but some words can rip-
cord out a raft that navigates the numerous gone.
And it could be either one would say

you might not want to put this stock,
these years, a career that's like that grape
that never got to be into poems. And I would listen.
But, you know, nobody that's really here—
all the many wrong ways of doing things
have always made a place for me.

ON THE GENERAL BEING OF LOSTNESS

Lostness is the *You Are Here*, the red star
that the mall map linked to GPS.
As if you'd stared into your nowhere
like a sun and photoreceptors
compensated with a point.

Lostness is an immaculately well-dressed
person or a room laid out like charcuterie.
It's a feeling someone loves you after
a ten-minute talk. Oh yes, but lostness
is loving someone too, knowing you would

take the raft out farther if it meant
a few more minutes. Sometimes,
I want to tell my dog I'm the only one
in the world who knows her whereabouts
and that's lostness but it's lived in.

It isn't sadness. Lostness is the job I had
In '98 in a warehouse unpacking chic decor
where I began to unravel and unmake
the very things the company was selling.
It was the boxes I moved forward

on the shelves until they lined up well,
pop choruses that played again for the beautiful
and found. It's almost gladness.
It's the walk I took one day trying to decide
Should I live in Montreal? and thinking that I knew

something that would make it plain.
Lostness is the many rains of money
that I once watched from an open window.
It's long been here. The semi-lunate carpal
flowering in late-Cretaceous bones; where everything

was going then never more unclear.
It was the first prokaryote closing off its little O
and all that it could be instead. But lostness is a steady wage.
I remember when my grandfather would come home
from the squats and thousand double-checks

of electrical work and wash his hands:
all the dirt moved in his laundry sink
like garter snakes that turned up under stones,
a living current so bearable in its lostness
that I could know it, only, for a hundred years

and still be happy. Lostness was the school
I went to where leaving crumbs on rectangles of paper
meant showing the someone would have to come.
It was having your knapsack up on the table
like a personal flotation device. It wouldn't be wrong

to say that lostness is always there on the lip of everything,
like lichen or a bomb. There is a loving lostness
that if you look deep into, you see a great
balance beam that everything
that was, or is, or that may be, is standing on.

DREAM OF DEE

From the casement window that looked out
over no particular sight, telephone lines
laid out a staff on which a starling would alight
like a one-beat rest, as if the wind, the branches
and, hell, all of the grid's wellspring, was frozen.

But it came back—I knew because the weird raccoon
would perform his evening stunt along that wire
holding whatever power we had braided there.
Once, I'd seen a pack scrimmaging through the lot's
untapped tabula rasa, some elsewhere shading

to in-the-know, knot shook free of itself
on some long-sunken dock. And once I had a dream
of you where you went under clear blue water
on a dock somewhere I've never been
and so I grabbed your collar while your eyes fixed

themselves in me, two worlds, the way a cure
and its own illness must. And I carried you ashore
to where in all this you just sprinted
beyond the world with me in it.

DRZYMALA'S WAGON

> Michael Drzymala is an early twentieth-century Polish folk hero known for exploiting a loophole in Prussian law that allowed him to live in a dwelling (a circus wagon) on land he owned in the province of Posen (Polish land that was considered a German territory). A policy set up by Otto von Bismarck meant to increase German land ownership (and drive out Poles) by making it difficult for Polish people to settle in this particular area, but Drzymala got around that by resourcefully occupying a moving dwelling (the wagon) on his land for years.

Drzymala, come and pick me up
on any so-called rehabilitated street
in Parkdale or Cabbagetown
riding that circus wagon you lived in for seven years
defying Prussian law in Kaisertu.

Take me around so I might see again
the loft spaces that the Ubisofts
will always have in the end, but often not the during,
and perhaps not even when the end itself
must pack up and go. Yes, I know,

we have what's ours and then it's not
what it was. A phone call or a door knock
has a slight transmogrifying force.
Or, for you, some officials fast approaching
as if sent out from the meeting place

of haze and horizon where an eye has a hard go
and the sunlight snows itself right out.
So I can't help calling up the sight of you
moving that wagon just a few feet every day
to get around the ban to build a dwelling on your own land.

You could turn and turn a life away.
Here, the salons all slide a Chablis into your hand
while you wait. Whatever we had,
it was never what it was I think I'd want to bet.
It was never wholly cordoned off from winters

where it can feel that all living things
are caught in someone's birthday candle blow.
That's the no that's ever sweeping in
like a cold front we can't see,
but in your mobile pad we stand outside of it—

or so inside it we are free. Around us
flags go up and down on a through-line
that stays hidden. I'd only hope that I could stay
with you, flagless, in the fantasy that turns
and turns toward the sorrow of the world.

DEAR LISTENER

Not reader, who had the page to confer
with and worry in or turn forever.
No, the one who sits, pupils rising up
like beer bubbles thinking what is this
exactly, happening in front of me,
like hand-spinning or blacksmithery?

Listener—who has the choice
to cut and run but who stays, quiet,
not quite with faith (but *faith*'s not the worst word for it)
and finds a feedback-swaddled room
inside this one in which another's voice
may play off opposites. As in this place
we've wound up in together.

Listener—a voice is calling out to you.
As if *you* were the one who hid
and wouldn't come out like a meaning.
Not because of what you feared
but reasons that you have yet to say
and will say yet. *You were here*
is a way to start, should you ever have a listener.

PLATYPUS

Like a freeze burn or a standing chair
it's all the categories throwing up

their hands into the air and saying, *Fuck it.*
All the grade-four projects flock.

Its mixed bag of not-too-classy tricks
is all that doesn't sit well with trophies,

curfews and the vagaries of pluralizing nouns.
There are those flags of algae that share

more genes with humans than they do with flowers.
They wave on shorelines like all the flags of countries

that never got to be. And those objects out beyond Neptune
that have the stones to demote planets

and then even begin the thankless task of adding more.
Is it really just friends' headshots

and their bullet-point accomplishments that make me
fall down again on *should* from *is*, *no* from *not*?

No—Gutenberg took all the screwy things from old
wine presses and then he never looked back.

What it is is a crock. In '68 Duane Allman
jammed "Hey Jude" with Wilson Pickett

who was reluctant at first but when they stepped outside the chorus, hey, there was Southern Rock.

GUITARIST

I spent years trying to be one.
Failed tuning in to most of high school
(and much beyond) because of it.
Seems I spent a basement-year
learning "Bensusan" by Michael Hedges,
who skidded on a rain-slicked curve
somewhere outside of San Francisco
and died in 1997. A digression
that sometimes makes me think
it's better not to follow tracks
with such sharp turns and giveaways
and giveouts that you can sense but never see
when young, and even a scrap heap shades
to a Wurlitzer. Problem, as ever,
the propensity to noodle,
though an added vice was the minor 7th
that when construed in reverb
shores up the shoegaze era in a squall.
What was I doing with all those years?
You can hear it in Bill Evans, too,
who's impressionism detonated in a breezy jazz
that never failed to open a window
even in winter and especially in rain
and, playing solo, his hands would go walking
out into Plainfield or Newark or wherever
he happened to be playing that day,
and might get an ice cream or abscond to a park
or keep going out of this altogether
and pick through the low cloud cover,
keeping a time we'll never know

but will also have to live by.
The whole big bazaar, which contains itself
like a window carries with it sky and people,
or as a chord contains the feeling
we're more seaward in our kitchens
than we'd want to bet. The long progression gives out—
but just slow enough to maybe think
that we're each separate, on our own notched tracks
with end points that wait for us, just
as the best songs seem to have been
plucked from air. Though dusk falls
and takes the colours back, and night
hikes the rent on being ephemeral.
All things go up in the lungbrain of stars
mixing and unmaking. It seems like some
invisible hand is ever modulating us
and, thus, Hedges dropped his lowest string
and thumbed it in his percussive fret-board
trances to anchor the diamond shatter of notes
and regain a centring force.
A chord's just a bit of the day put to use,
whose making defines not an infinite plain
but the field in which one might re-contact
all that's assumed to have returned to its state—
now, I play the laundry drying rack's 30-gauge
on afternoons where nobody waits for
either of us on an un-gated stretch of Bloor.
I hang the clothes we both wear
for the years we have left. There's
everything the brand has washed away from
and all that's branded by its wear,
which is the sweater that only you've seen me in
or in another case a cardigan, which given

the right time and place, removes the chill
without you breaking a sweat. Oh yes,
I wanted to be something. Only happy
after many years that I never was.

THE JOEYS AT KANGAROO CREEK FARM

They passed one on to me in a sanctuary
in Lake Country, little schooner
on a stream of hands. My arm, an instant valley
as if the newly born expanded what was really there
abruptly and so maybe really wasn't,
as in *there*, I mean. You were I think in Saint-Denis
but could have moved back to Davis
following those quantum paths
I always just saw as a kind of bland Sudoku.
And what would you name it? someone asked,
and I knew no other living thing
had been so named, so filled up with all the many sounds
it wasn't, and I liked the quiet so I
just thought I'd name it the first thing that came:
Shockmaster—after Fred Ottman's scrambling
in that so-awful-it's-a-masterpiece of something
clip from '93; the unscripted flailing
that sent his helmet rolling outwards
across the prime Daytona studio
like Pluto did and maybe Britain
all slinking away from the same plaster-patched-together thing.
And might as well add us there too,
hell, it's happy hour, which is maybe
what the joeys do to anyone.
A masterpiece of falling is what I would say
it maybe was. The joey was a masterpiece
of being pulled up from the reeds
that grow just on the edges of not being
and that I think we spend our lives somehow
so tangled in. A thinking reed is what Simone

de Beauvoir says we are. Of course including me
who was still pulled up now so long ago
it's just a dream, so far gone it might as well
have been in Delaware for all I knew.
So after everything, there was nothing else
I could think to do but pass it onwards.

ONLY AN AVENUE

To have known it then, to have chosen,
hearing expressways shush above it,
the sorting line of interstates and viaducts
where the perseverance of water is managed.
Good, clean city work, hours funnelling
down the sluice and asphalt nodding
off to a deep, unflinching sleep
while each of us comes to something no matter
how far down on the company ladder we lean.

Turning onto it, its many signs of play
and warning, its oaks, spruces, redwoods,
perhaps lined equidistantly, and city life,
at night a thousand cabins lit and moving
closer to the speed of light than we would have guessed.
And avenues, like rivers, keep continuing;
they are rivers stilled and never leading
to an ocean, water slowed down and wave-less.
They go where they go as long as it's possible.

They are the work of all who came before
this but are gone, so gone that it must
be what peace means. It's a nothing
that can fold you in or unfold in you
the handful of things you hold when free.
And you don't ruin it when you speak.
And we don't compromise it with position.
To know that and go on down and through
I'm not sure where, and say that, yes,
I want to be. This one way, too.

OATH OF AN UNAFFILIATED BOY SCOUT

To know that no one and nothing is coming.

To fall asleep outside in a light rain.

To find peace streaming in through a far open window.

To the hi-hat and kick in a storm, to the light show.

To know a foreign place, like New Jersey, by heart.

To move always beside the river of daughters.

To live well in the smallest cubic space.

To never put out a lard fire with water.

To be all right with nobody's directions.

To hold in place all the days we weren't here.

To know it will take many years but might not.

To somebody's nice point never being the last thought.

To be there for God should God be there and doubt it.

To shout it, sometimes—and others to still shout it.

To under-read enough to have friends.

To over-read enough to be moral.

To know there's no bedrock but still agree.

To roam far while still holding someone's hand.

To live, for as long as you can, in the difficulty.

THE GREAT ILLUSION

We're just sea slugs. More or less Silly Putty
pulled across a hot mess of fold and bone.
Have you seen a brain? It's a labyrinth of looking.
One minute, you're in a tunnel of neuronal swag
never cashing you out it seems, next you're in

the rarified air of thought. You could call it dream.
I think of those contortionists scrunched up
in the halved saw box that a magician wheels.
There's some flashy trick to being—some sleight
of sheer anatomy the naked mind's eye can't reveal.

Speaking of the fits that looking closer brings,
in '83, Benjamin Libet discovered that a person
could feel they're making a choice when they were not.
A little dot that rolled around on a screen
and someone pressed when they wished. That was it.

Our inner lives were a great illusion. Twenty-eight years later,
Alex Rosenberg wrote *The Atheist's Guide To Reality*
a book that followed science out into the parking lot
and then got in the car. Our everyday sense of being
evicted from the real and true for a few electric shivers.

One summer, I sat in a café and read that book and I knew
less than ever about what to do. Put simply,
I was heavily despairing beside a dog named Captain.
No need for a story or poem's dense, remaindered weight;
no aching that a Celexa could not un-rain.

I had dreams of being wheeled out to a cliff and tipped
over out onto the for-some-reason-always-European peninsula.
Then the wheelchair was calmly pushed back.
It wasn't that I didn't know we're all ultimately
playing cricket in the dark. Take anything—the sky at night;

those flies that show up under streetlights
in traceless swirls of coming and going, that swervy,
strained invertebrate flight that pings there like electrons
and is gone when you shine the brighter thing—
all the hinges in common sense break.

And Rosenberg's next book turned out to be a novel.
I remember thinking this one day my father and I
were driving the MG that he made from parts.
We'd left a rose for my grandfather.
And a gravestone is the only place your name

will weigh a hundred pounds for more than an evening
I think I said. He was telling me about a time he thought
he saw Bono in a Home Hardware and then assumed
it was an impersonator, but really it was just somebody
walking around in really big sunglasses.

And Rosenberg's next book turned out to be a novel.
Maybe all of what we thought was really in place
was just a lure that bobbled while we followed it.
Could it all boil down to such a slip
of judgment in the ouroboros true?

Later, we went over that one dip in the expressway
where it feels like you're falling off the Earth.
Maybe we were. And this thought occurred,

just floated up into the reservoir I wasn't:
It was meant to prove our wrongness

 in a way that felt less microfossil,
but Rosenberg's next book turned out to be a novel.
My brain kept me baffled but still buoyant
in that semi-light. Conscious life was a treading
in the condo-coral strewn oblivion.

It was hewn in that oblivion, too. What came up,
with luck, was a lifting bag in the brightness changing hue.
A thought was a craft I had in the night.
And then the night was what I had in lieu
of all the ways that it was possible to say.

TWO CELLS MADE ALL OF THIS

I'm sitting now outside of all I've said
and heading south on the I-90, which makes clear
the vast stretches of the way a life can go
and in this brand of dusk all seems power-washed
of time and place, and I've been watching both
the signs for deer and then, sometimes,
the ones that were seen all too well, muscled abruptly
out of car-shot and left to lie looking up
as if dumbstruck by some huge enormity
the sky keeps behind itself, like the "actual" sun.
The trees here are gathered like some poker hand
that always wins, and, then, as if between blinks,
a town that never had any business being
makes its business in the day and often in the dark.
I forgot to say I'm heading to the Dickinson
house and the many parking garages of Boston.
There's a name now for what she suffered from
and had to lie down looking up as rooms would spin
above her in some terrible realization of her seeing's flex.
Sometimes a complication's good. She kept her work
in fascicles and each time a new edition condenses,
it seems a final take just rains itself right out,
a fact you can intuit in a time where there's an
ultraviolet view to see what's there when the sliver
of visible spectrum widens to an un-tuned forever.
It's good at times to think of this, though good
to just be taken and so in this outer stretch I am—
in deep, that is. Some days there will be a clarifying
of the predicament. I know once all this was just
a drafty stretch of eon space on which the raftless

and the ad hoc sunk though one extra complication
granted me a kind of perch. And so I'm here.
I'm thirty-five for one more week on the interstate.

THE FLY

To say how I merged on it would be to say
how I got anywhere. Bottom line, I was

on a highway that sloped upwards out in front of me
for an ever that seemed up for grabs.

I stopped and bought my meals on credit
that was always given, and everything was wrapped

in cellophane and tasted like Knorr.
I heard the diners, making out a word or phrase

but mostly couldn't. Oh, and the strangest part:
it seemed each mile I drove rewound the time

I'd put behind me like I was travelling backwards
into my life. Which meant that whatever exit taken

found me visiting a place I'd lived once for a moment
and all I'd known there aged but in reverse.

Funny, though, that even if big moments were apparent,
it mostly wasn't watchable. Seems what happened

had been logged in some cache the world was keeping,
but so much was some minutes just obstinately there

coming up out of the darkness of the heating vents
or soaking your shins from the sprinkler heads of years.

Things went like this until I said, *It's not my life,*
whatever's here is the universe's staging of a punishment,

and I called out to all the dwellers there but everything
I vocalized came out as if underwater.

Attempting to throw glasses and thus go poltergeist on the fracas
fell on deaf ears the world had long perfected.

Outside into the shared backyard, I was thinking
it was all really just the same now, no?—all alone

among lawns the corporation kept so green and glistening?
And I was thinking that when a fly landed on my wrist.

It polished all the stained glass of its wings and I could see
the mix of shell and hair it was, its eyes like two pebbles

now given flight and sight and its legs like twigs
now fused with sound. Then it was gone.

And I swear that when I looked up I could hear everything
around me thrumming, the cicadas and the crickets

and the hydro wires and then the neighbours clanging cutlery
talking of the storm that came, the strumming of a lone guitar,

somebody crying and then whispers I was following
through windows with a radio and the humming of a woman

in a rocking chair. And just like that night had fallen everywhere.
You'd think the sound would have died down but it didn't.

DREAMPAD

> *Just wanted to let you know we are extremely pleased with the Dreampad.*
> *We now call it the "magic pillow" and my son really likes listening to the music.*
> —Review of the Dreampad pillow

At night, I lay my head down on the slow
obliterating sound of hoof beats in the Ozarks.
The Chinook of my looking blows through many streets.
When I'm carried through country after country

to the tune of work, there is a disappearing
into keyboard taps and singular cash register riffs.
There's a static on our shorelines that we might
turn up. My Dreampad emits life-music

that my body can conduct with its boutique
antennas cued. Might be that it's above me.
But when I'm laying supine it just feels kind of lovely,
an IV to everything via one simple USB.

Let me be just. Let me be free. That there is a deep green
exploding outwards is enough. My dreampad is a buffering
of waking-ness, of wanting, and so it takes the suffering
I might be and stretches it again, again, into a synth note.

If justice means rebut me, then upend my life.
My best case would leave too many screwed.
If there's a path to find my way back to that one sand dune
I think of as the realest place I've ever been

then please let me. One day I'll go into the feed forever,
and there will never be a way to find the Earth again.
So each night before I sleep, I say, *No matter me,
no matter you.* I think there is a kind of rain

that can fall again, in the end but not finally,
because it's never through. *No matter me, no matter you.*
Because there is another way beyond
the way it is, that it could still even be.

NOTES ON THE POEMS

I reference a poem called "Stargazing" by Glyn Maxwell in "Komatsu Floodlight."

The epigraph used in "Sky Pool" is an interview quote taken from an article from the Verge called "London's sky pool will let the super-rich swim through the air" by James Vincent.

The poem "A Mile From the Bay of Biscay On Tour With Oneohtrix Point Never" is entirely made up but does follow a plausible tour route Lopatin might have taken and includes some biographical detail gleaned from interviews, which are the make of his synthesizer and places he's lived. All of the rest of the poem is fictional.

"Spacetime" adapts details from a *Guardian* article called "The Journey" by Patrick Kingsley.

The epigraph used in "The Bright Note" is taken from Douglas Coupland's article "A radical pessimist's guide to the next 10 years" from the *Globe and Mail*.

"Growth" begins with a similar concept as does Glyn Maxwell's poem "Thinking: Earth."

The epigraph used in "Clear Giant" is transcribed from the Radiolab episode "Forget about Blame?"

The epigraph used in the final poem, "Dreampad," is a customer review of the Dreampad pillow from the Dreampad Reviews website.

ACKNOWLEDGEMENTS

Poems from this collection appeared in *New Poetry*, *The Walrus*, *The Puritan*, *Poetry Magazine*, and the 2016 edition of *The Best Canadian Poetry*. My deepest thanks to the editors of these publications.

Poems from this work appeared in a chapbook published by Anstruther Press called *Helium Ear*. My thanks to Jim Johnstone and Erica Smith as well as the Anstruther team.

Thanks to the wonderful team at M & S: The poetry board, Anita Chong, Kelly Joseph, and everyone else that contributed to design, layout, typesetting, and marketing of the book. Rachel Cooper created an incredible cover design. Deepest thanks to everyone.

Thanks to David Brock and Paul Vermeersch for being those readers you really need in order to straighten out, and then double down on, the strangeness a poem requires to have whatever liftoff it might achieve.

Thanks to numerous friends who have looked at drafts of my work.

Thanks of course to Kevin Connolly for his exquisite editorial eye.

And, as always, to Holly Kent. We have this joke that when I approach with a piece of paper, I also hum the music from *Jaws*. Whatever is here, though, is due as well to her willingness to be a reader in the late and early hours and to come along on this little skiff of a writing life. Blessed be the cruise ships, but I've stopped looking out at them.

The text of *Dreampad* has been set in Stempel Garamond, a modern font family based on the types of the sixteenth-century typographer and printer Claude Garamond (who had modeled his types on those of Venetian printers of the previous century.) The Stempel foundry based its version—featuring a more angular and incised appearance, and a slightly heavier weight—on a sample sheet of Garamond's original type. Stempel Garamond was released in 1924.